Cover and internal design © 2020 by Sourcebooks

Cover and internal design by Will Riley

Internal images © by Terdpongvector/Freepik, macrovector/Freepik,vecteezy.com

Sourcebooks and the colophon are registered trademarks of Sourcebooks.
Published by Sourcebooks eXplore, an imprint of Sourcebooks Kids

P.O. Box 4410, Naperville, Illinois 60567-4410

(630) 961-3900

sourcebookskids.com

First published as Red Kangaroo's Thousands Physics Whys: *Going Back in Time: General Relativity*
in 2018 in China by China Children's Press and Publication Group.

Library of Congress Cataloging-in-Publication Data is on file with the publisher.

Source of Production: PrintPlus Limited, Shenzhen, Guangdong Province, China

Date of Production: June 2020

Run Number: 5018701

Printed and bound in China.

PP 10 9 8 7 6 5 4 3 2 1

Let's Time Travel!

Zooming into the Science of Space-Time with General Relativity

sourcebooks
eXplore

**#1 Bestselling
Science Author for Kids
Chris Ferrie**

"Ugh! My tummy really hurts," Red Kangaroo groans.

"I probably should have eaten more vegetables instead of all that junk food. Maybe Dr. Chris can help me travel back in time so I can feel better!"

Red Kangaroo makes her way to the lab to find Dr. Chris. "Can you tell me if it is possible to go back in time?" she asks him.

"Time travel actually might be possible!" says Dr. Chris. "You just need to understand Einstein's theory of general relativity!"

"I think I've heard of Einstein before. Wasn't he a famous physicist?"

"That's right!" Dr. Chris replies. "**Albert Einstein** came up with theories about space and time. His theory of **general relativity** can help us think about time travel because it is all about curved space-time."

"Curved space-time?" asks Red Kangaroo.
"How can time and space curve?"

"Thinking about space-time confuses even the greatest scientists," Dr. Chris says. "Why don't we start with an easier idea first, like curved paper?"

"Sounds like plan, Dr. Chris!" Red Kangaroo says.

"There are four **dimensions** in space-time," says Dr. Chris. "Can you give me an example of something with one dimension?"

"I know! I know!" cries Red Kangaroo. "One dimension is a straight line. Its only dimension is length. Kind of like this piece of string!"

"I think two dimensions would be like a sheet of paper, because it has both length and width," says Red Kangaroo.

"That's right!" Dr. Chris replies. "Now there are just two more dimensions to go."

"The first and second dimensions are easy!" says Red Kangaroo. "But I don't think I know the third dimension. What does that look like, Dr. Chris?"

"It's just as easy," Dr. Chris says. "We live in three dimensions. This room has three dimensions, and our bodies take up three dimensions of **space** because they have length, width, and depth!"

"Wow! I've lived in the third dimension my whole life and never knew!" says Red Kangaroo.

"But wait," Red Kangaroo says.
"You said there were four dimensions.
Is the fourth dimension easy too?"

"Yes, it is!" Dr. Chris says. "You're just as familiar with the fourth dimension as you are with the other three, because the fourth dimension is **time**!" says Dr. Chris.

"Wow! That is easy!" says Red Kangaroo. "Three space dimensions and one time dimension. I can remember that!"

"I can see examples of the other dimensions, Dr. Chris," Red Kangaroo continues. "But I can't see time! How can I understand the fourth dimension if I can't see it? And what does this have to do with time travel?"

"I have an idea!" Dr. Chris says. "Let's imagine all of space and time is this sheet of paper."

Past ——————————

TIME

"One end of the paper is the past. The other end is the present, or right now," says Dr. Chris. "The space doesn't change but time does when you moved from one end of the paper to the other."

→ **Present**

Dr. Chris

Past ———————————

TIME

"So all I need to do is go to the beginning of the page and I'll be in the past again!" Red Kangaroo says. "But wait...how do I go back in time if it's always moving forward?"

→ Present

Dr. Chris

"All we need to do is bend the fourth dimension," Dr. Chris says. "By curving **space-time**, you can put the past on top of the present. And then you would have traveled back in time!"

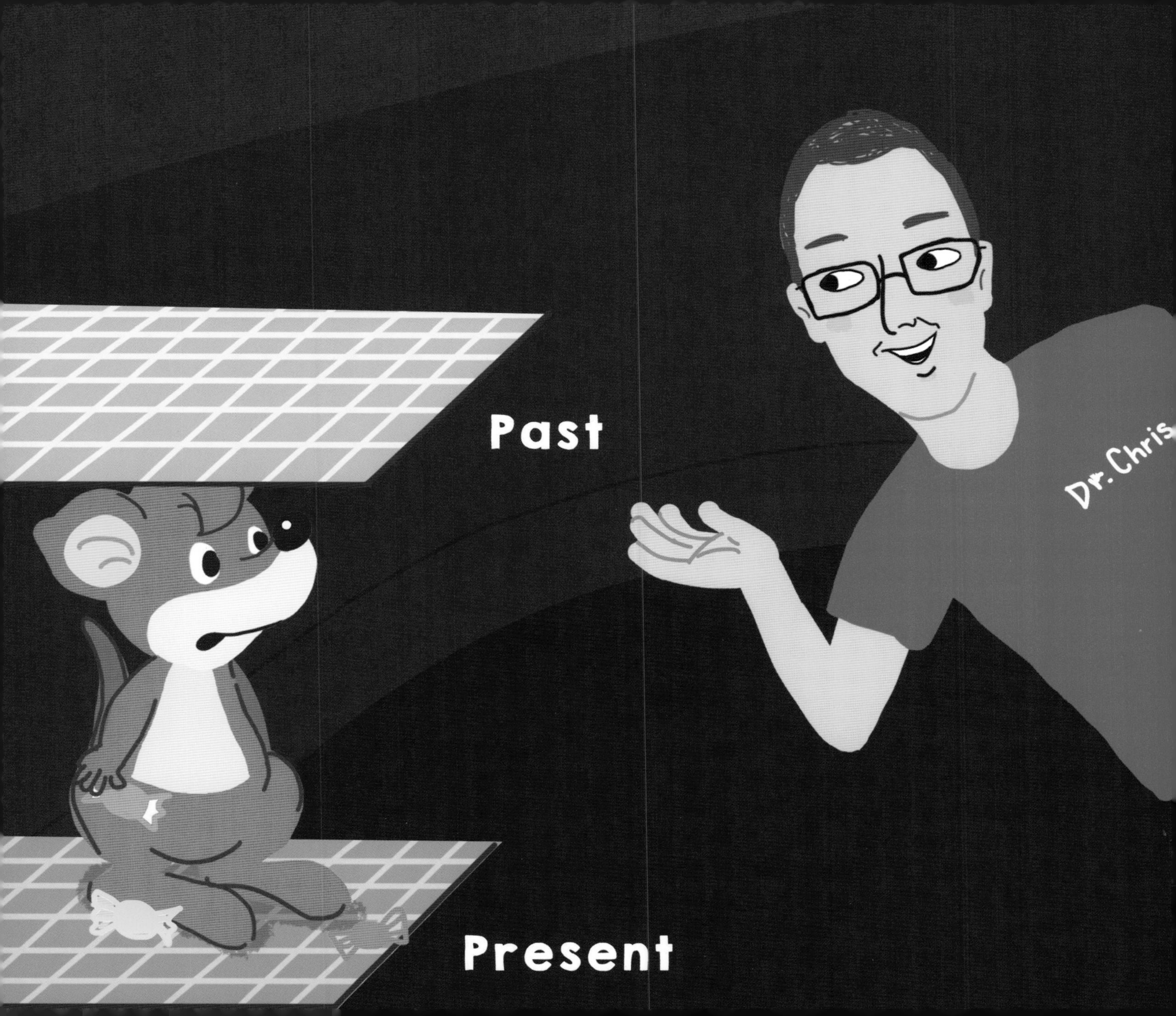
Past
Dr. Chris
Present

"That sounds so easy!" Red Kangaroo says. "I want to time travel! Can we bend time right now?"

"Not so fast, Red Kangaroo!" Dr. Chris says. "Scientists haven't actually figured out how to bend space-time like this. But maybe you will be the first to figure it out!"

"I can't wait to figure out how to bend space-time," Red Kangaroo says. "But first I need some food. And this time I am going to eat healthier since I can't travel back in time just yet!"

"Okay! I feel much better now, Dr. Chris! Let's get to work!" Red Kangaroo says.

"Great!" Dr. Chris says. "Are you ready to play with the four dimensions of space-time?"

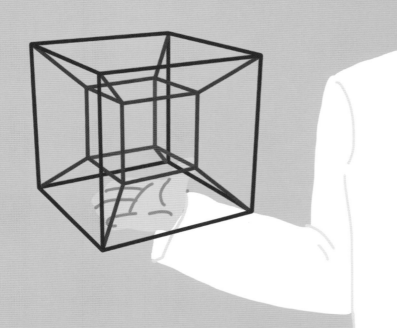

Glossary

Albert Einstein

(1879–1955) A German scientist who created many theories of physics, including relativity and quantum physics.

Dimension

The amount of numbers needed to locate a point in (or on) an object. There's one dimension for a line, two dimensions for a sheet of paper, and so on.

Space

The three dimensions in which all things exist and move: length, width, and height.

Time

The flow of events from the past, through the present, and to the future.

Show What You Know

1. How many dimensions does space have? How many does time have? And space-time?

2. Count the number of dimensions of space that the words on this page exist in.

3. Count the number of dimensions of space the page itself exists in.

4. This page does not need to be flat—how can you show that it can also be curved? What other things around you can be either flat or curved?

5. Name the scientist who invented the idea of curved space-time.

Answers on the last page.

Test It Out

Warping geometry

1. You will need paper, a pencil, a large ball (a basketball, a soccer ball, or even a globe will do), tape, string, and a protractor.

2. With the straight edge of the protractor, draw many different-shaped triangles on the piece of paper.

3. Make a prediction: Will any of the triangles have the same total degree of angles?

4. Use the protractor to measure the angles of each triangle. Add up the numbers in each to figure out the total degree of angles each triangle has. Was your prediction correct? Were any of them the same or were they all different?

5. Now use the string to help you create a triangle along the surface of the ball. Use the tape to hold the string in place.

6. Make a prediction: If you added up all the angles on this triangle, will the answer be the same as any of the triangles you drew and measured on the paper?

7. Now measure the angles of the triangle on the ball and add them up. What did you discover?

8. Try this experiment with other curved surfaces. Make predictions and discover what the angles of a triangle will add up to on them!

Making space-time

1. You will need a large bowl, plastic food wrap, and some different-sized marbles.

2. Stretch the plastic wrap over the bowl, but not too tightly.

3. Make a prediction about what you think will happen to the plastic wrap if you put the largest marble on top of it, right in the center. Now place the largest marble in the center of the wrap and record what happens. Did it match your prediction?

4. Make a prediction about what will happen if you add a smaller marble on the plastic wrap near the edge of the bowl. Gently place the smaller marble near the edge and record what happens. Did it match your prediction?

5. Now give the smaller marble a little push so that it will travel along the edge of the bowl. What does this remind you of?

6. Play! Try sending the marbles around in different directions, or using an even heavier or lighter object in the center. Think about what is causing the marbles to move the way they do.

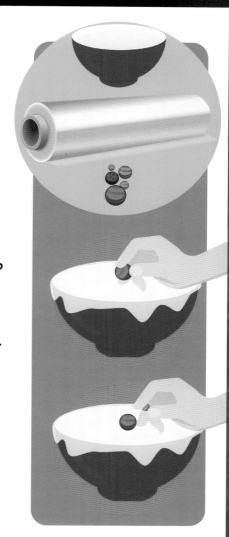

What to expect when you Test It Out

Warping geometry

On flat surfaces, all triangles have angles that add up to 180°. But geometry is different on curved surfaces. For a sphere, like a ball or the Earth, the angles add up to more than 180°. Other shapes, like a saddle or a donut (also known as a torus), have angles that add up to less than 180°.

Making space-time

The plastic wrap is like space itself. When you place the heavy object in the center, that object acts like a lot of mass that warps space around it. You can think of it like the Sun. A small object, like the little marble, does not warp space as much. It is attracted to the larger object because it feels the warping of the space. Sending the marble around creates a little solar system, like Earth revolving around the Sun. The marble eventually falls into the center because of friction. Earth does not experience much friction, but it is slowly losing energy and moving closer to the Sun all the time. (But don't worry, it'll be a really long time before the Earth and Sun meet.)

Show What You Know answers

1. Space has three dimensions and time has one. Space-time is the combination of those four dimensions.

2. The words on the page exist in two dimensions of space.

3. The page is part of the book, which exists in all three dimensions of space. This means you can move it anywhere you like!